Epigenetic Longevity

How Lifestyle, Nutrition, and Environment Reprogram Your Genes for a Longer Life

By Joe Correa

Published by Live Stronger Faster

First Edition

Table of Contents

Introduction

For decades, aging was framed as an unavoidable consequence of genetics—a slow, predetermined decline written into your DNA at birth. If longevity ran in your family, you were lucky. If it didn't, the narrative suggested that disease, fatigue, cognitive decline, and shortened lifespan were simply a matter of time. This belief shaped medicine, public health, and even personal responsibility for aging. Yet modern science has decisively overturned this outdated model. We now understand that genes do not act in isolation. They respond. They adapt. And most importantly, they can be *reprogrammed.*

This is the domain of **epigenetics**—the science of how behaviors, environments, and internal signals regulate gene expression without changing the underlying DNA sequence. Epigenetics explains why identical twins age differently, why lifestyle choices can accelerate or slow biological aging, and why interventions made later in life can still meaningfully extend healthspan and lifespan. In the context of longevity science, epigenetics represents one of the most powerful leverage points available to human beings today.

This book exists to translate epigenetic science into actionable longevity strategies. While previous volumes in this series addressed cellular repair, mitochondrial energy, hormones, inflammation, nutrition, sleep, and brain health, *Epigenetic Longevity* ties these systems together at the level where they ultimately converge: gene expression. Every habit you perform, every nutrient you consume, every hour you sleep or miss, every stressor you tolerate, and every toxin you encounter sends biochemical signals to your genome. These signals determine which genes are turned on, which are silenced, and how efficiently your body repairs, regenerates, and protects itself over time.

Aging, therefore, is not simply the passage of years—it is the cumulative expression of epigenetic signals.

At the core of epigenetic regulation are molecular mechanisms such as DNA methylation, histone modification, chromatin remodeling, and non-coding RNA signaling. While these terms may sound technical, their functional impact is straightforward: they act as switches and dimmers on your genetic blueprint. When longevity-promoting genes are expressed efficiently, cells maintain resilience, inflammation stays controlled, repair mechanisms remain active, and metabolic flexibility is preserved. When epigenetic signaling becomes distorted—through chronic stress, poor nutrition, environmental toxins, sleep disruption, or sedentary living—aging accelerates, disease risk rises, and recovery capacity declines. One of the most profound insights to emerge from epigenetic research is that **biological age is malleable**. Unlike chronological age, biological age reflects the functional state of your cells, tissues, and systems. Epigenetic clocks—tools that measure DNA methylation patterns—have demonstrated that individuals of the same chronological age can differ dramatically in biological aging. More importantly, these clocks have shown that targeted lifestyle interventions can *slow, halt, or even partially reverse* biological aging markers.

This means longevity is not merely about living longer—it is about maintaining youthful gene expression patterns for as long as possible.

In this book, you will learn how lifestyle inputs act as epigenetic instructions. Physical activity activates genes involved in mitochondrial biogenesis, antioxidant defense, and insulin sensitivity. Strategic fasting triggers epigenetic programs associated with autophagy, stress resistance, and cellular renewal. Nutrient-dense diets supply methyl donors and cofactors essential for proper DNA regulation. Sleep synchronizes circadian genes that govern repair, detoxification, and hormone balance. Stress management prevents the chronic activation of genes linked to inflammation, immune dysfunction, and accelerated aging. Equally important, you will learn how environmental exposures—from air quality and water purity to light exposure and chemical load—interact with your genome. Modern environments expose humans to epigenetic disruptors that previous generations never encountered at scale. Understanding how to mitigate these influences is not optional for longevity; it is foundational.

This introduction serves three critical purposes. First, it reframes aging as a **dynamic biological process** rather than a fixed destiny. Second, it establishes epigenetics as the *control layer* that integrates every longevity intervention discussed across the entire 10-book series. Third, it prepares you to think systemically—recognizing that isolated interventions are less powerful than coordinated epigenetic strategies applied consistently over time.

Unlike superficial anti-aging approaches that focus on appearance or short-term performance, epigenetic longevity operates on deep biological timelines. Changes at the gene expression level compound quietly but profoundly. Small, repeated signals—daily movement, regular fasting windows, consistent sleep routines, controlled inflammation—accumulate into durable physiological advantages. Over months and years, these signals reshape how your body allocates energy, repairs damage, and resists degeneration. The goal is not perfection, but consistency—creating an internal environment that continuously nudges your genome toward resilience rather than decline.

You will also learn why epigenetic aging accelerates during midlife and how this period represents both a risk and an opportunity. Many individuals experience a rapid shift in metabolic health, inflammation, and hormonal stability between their late 30s and 50s. Epigenetically, this phase reflects accumulated signaling errors—but it is also the window where corrective strategies produce outsized benefits. Intervening at the epigenetic level during this stage can dramatically alter the trajectory of aging for decades to come.

Importantly, epigenetic longevity is not about extreme biohacking or experimental interventions reserved for elite laboratories. While advanced technologies and supplements may enhance outcomes, the foundation remains accessible and behavioral. Your genes evolved to respond to movement, nutrient scarcity and abundance cycles, light-dark rhythms, social connection, and environmental feedback. Modern longevity science is, in many ways, about restoring these signals in a controlled, intelligent manner.

As you progress through this book, you will encounter recurring themes: adaptability, hormesis, signaling balance, and biological feedback loops. These concepts are not abstract—they are the language your genome understands. By learning to speak this language fluently, you gain the ability to influence aging at its source.

Epigenetic Longevity is not meant to stand alone. It is designed to interlock with the preceding and following volumes in this series. Cellular rejuvenation, mitochondrial health, hormone balance, inflammation control, nutrition, sleep, and brain function all exert their effects through epigenetic pathways. This book explains *how* and *why* those interventions work—and how to optimize them for maximum lifespan impact.

Ultimately, longevity is not a single decision but a continuous conversation between you and your genes. Every day, that conversation is happening whether you are aware of it or not. This book gives you the knowledge to participate intentionally—to send clearer signals, reduce noise, and reinforce the biological instructions that favor long life, sustained energy, and preserved vitality.

A longer life is not guaranteed. But a *better-regulated genome* is achievable. And from that foundation, longevity becomes less about chance and more about design.

This is where that design begins.

Chapter 1

Your Genes Are Listening

From the moment you wake up, your genes begin responding to the signals you provide. Before you eat, before you move, before you consciously decide how your day will unfold, molecular instructions are already traveling through your cells. Hormones, neurotransmitters, metabolites, and inflammatory markers act as messengers, binding to receptors and influencing how your DNA is expressed. This process is continuous and relentless. There is no neutral state. Every internal and external input either reinforces resilience or accelerates decline.

Epigenetics explains why genes behave less like fixed blueprints and more like responsive software. The DNA sequence itself rarely changes, but the way it is read changes constantly. Chemical tags attach to DNA strands and histone proteins, altering access to genetic instructions. When genes associated with repair, detoxification, metabolism, and stress resistance are accessible, cells function efficiently. When those same genes are silenced—often due to chronic stress, inflammation, or metabolic dysfunction—aging accelerates from the inside out. One of the most important revelations in longevity science is that epigenetic signaling responds more strongly to *patterns* than to isolated events. A single unhealthy meal does not age you. A single night of poor sleep does not collapse your biology. What shapes gene expression is repetition. Signals repeated daily become biological expectations. Over time, your genome adapts to what it believes is your environment. If that environment is chaotic, inflammatory, sedentary, or sleep-deprived, your genes optimize for survival at the expense of longevity. If that environment is structured, nutrient-rich, physically active, and rhythm-aligned, your genes shift toward maintenance, efficiency, and long-term preservation.

DNA methylation is one of the most studied epigenetic mechanisms in aging. Methyl groups attach to specific regions of DNA and act as on-off switches for gene activity. With age, methylation patterns drift. Genes that should remain active become silenced, while genes associated with inflammation and disease may become overexpressed. This drift is not random. It is strongly influenced by nutrition, stress hormones, toxin exposure, and metabolic health. Research using epigenetic clocks has shown that individuals exposed to chronic psychological stress can exhibit accelerated biological aging equivalent to several additional years within a relatively short time span.

Equally important are histone modifications, which determine how tightly DNA is wrapped around histone proteins. When DNA is tightly packed, gene access is restricted. When it is loosely packed, gene expression increases. Physical activity, fasting, temperature exposure, and circadian alignment all influence histone structure. This is one reason movement and metabolic challenges have such profound longevity effects. They do not merely burn calories or improve fitness; they mechanically alter how your genome is accessed.

The body evolved to interpret certain stressors as signals for adaptation rather than damage. This concept, known as hormesis, is central to epigenetic longevity. Short, controlled stress—exercise, fasting, cold exposure, heat exposure—activates genes involved in repair, antioxidant defense, mitochondrial renewal, and autophagy. Chronic, uncontrolled stress—sleep deprivation, emotional distress, nutrient deficiency, toxic exposure—activates genes associated with inflammation, immune suppression, and cellular aging. The difference lies not in stress itself, but in duration, intensity, and recovery.

Modern life disrupts epigenetic signaling in subtle but cumulative ways. Artificial light alters circadian gene expression. Ultra-processed foods distort insulin signaling and methylation pathways. Sedentary behavior silences genes responsible for metabolic flexibility. Chronic psychological stress elevates cortisol, which directly alters gene expression in immune cells and accelerates inflammatory aging. These influences rarely cause immediate illness, which is why they are underestimated. Instead, they quietly rewrite genetic expectations year after year.

One of the most empowering aspects of epigenetics is reversibility. Unlike genetic mutations, epigenetic marks are dynamic. Studies have shown measurable improvements in epigenetic age following structured lifestyle interventions involving diet, exercise, sleep optimization, and stress reduction. These changes do not require extreme measures. They require alignment. When signals become coherent and predictable, gene expression stabilizes in a more youthful configuration.

This chapter establishes a critical principle that underpins the entire book: your body does not age because time passes; it ages because information accumulates. That information is biochemical, hormonal, and environmental. Epigenetic longevity is the practice of managing that information intentionally. It is not about forcing youth but about removing the signals that instruct decay.

As you move through this book, every protocol—nutritional,

environmental, behavioral—should be evaluated through one question: what message does this send to my genes? When you understand that your genome is constantly listening, longevity becomes less mysterious and more methodical. Aging stops being something that happens to you and becomes something you actively shape.

Daniel turns 47

When Daniel turned forty-seven, nothing appeared wrong on paper. His bloodwork hovered just inside normal ranges. His weight hadn't changed much in years. His career was stable, his family healthy, and his routines predictable. Yet something felt off. Recovery took longer. Energy faded earlier in the day. Minor injuries lingered. Sleep felt lighter, less restorative. Doctors offered reassurance. "This is normal aging," they said. Daniel accepted the explanation—until he learned what aging actually meant.

He discovered epigenetics by accident, late one night, searching for answers he couldn't articulate. What struck him wasn't fear, but relief. The idea that his decline wasn't inevitable but informational changed how he saw his body. For the first time, aging felt negotiable. Not reversible in a dramatic sense, but responsive. He wasn't broken; he was being misinformed.

Daniel didn't overhaul his life overnight. He changed small things first. Morning light replaced scrolling. Evening meals moved earlier. He walked daily, not for exercise, but for rhythm. Meals became simpler, more consistent. Sleep became protected, not sacrificed. None of these changes felt revolutionary. That was the point. He wasn't chasing youth; he was restoring coherence.

Within months, subtle shifts accumulated. Energy stabilized. Inflammation quieted. Sleep deepened. His body responded not because of a single intervention, but because the signals stopped contradicting each other. Stress hormones no longer competed with repair hormones. Blood sugar spikes softened. Circadian genes regained rhythm. His genome adapted to the new expectations he set.

What surprised Daniel most was not physical improvement, but psychological clarity. When his body stopped operating in crisis mode, his mind followed. Chronic low-grade stress lifted. Decision-making sharpened. Motivation returned—not as urgency, but as calm momentum. He realized how much of what he had accepted as aging was actually biological confusion.

Years later, Daniel looked younger than his peers, but more importantly, he felt *durable*. He recovered faster. He adapted better. He trusted his body again. Aging didn't disappear, but it slowed into something manageable, even graceful. He wasn't trying to live forever. He was trying to live well for as long as possible. Daniel's story is not exceptional. It is representative. Epigenetic longevity does not require perfect genetics, unlimited resources, or radical interventions. It requires awareness and consistency. It requires understanding that your body believes what you repeatedly tell it. When those messages align with resilience, your genes respond accordingly.

This chapter is not about optimism. It is about agency. Your genome is not passive. It is responsive, adaptive, and deeply sensitive to your daily environment. Every choice you make—how you eat, move, sleep, and recover—becomes a biological instruction. Over time, those instructions become identity.

Longevity begins the moment you stop treating aging as fate and start treating it as feedback. Your genes are listening. And they are ready to respond.

Chapter 2

The Epigenetic Switches That Control Aging

Aging does not occur uniformly across the body, nor does it progress at a constant rate. Some tissues deteriorate rapidly while others remain resilient. Some individuals age biologically faster than their chronological peers despite similar genetics. This uneven progression puzzled scientists for decades until epigenetics revealed the underlying control system. Aging is governed not by time alone, but by regulatory switches embedded within the genome that determine how cells respond to stress, damage, and repair demands.

At the center of epigenetic regulation are DNA methylation patterns. These chemical tags attach primarily to cytosine bases along the DNA strand and influence whether a gene is active or silent. In youth, methylation patterns are precise and tightly regulated, ensuring that genes required for growth, repair, detoxification, and metabolic balance remain accessible. With age, these patterns drift. Genes critical for cellular maintenance may become increasingly silenced, while genes associated with inflammation, fibrosis, and metabolic dysfunction become overexpressed. This phenomenon, known as epigenetic drift, is one of the most reliable biological markers of aging.

Epigenetic clocks leverage methylation patterns to estimate biological age with remarkable accuracy. These clocks reveal that lifestyle factors such as chronic stress, obesity, smoking, poor sleep, and metabolic disease accelerate epigenetic aging, while structured exercise, caloric regulation, and stress reduction slow it. Importantly, these changes are not purely cumulative. Epigenetic marks respond dynamically to improvements in signaling environments. This means that aging acceleration is not permanent, but conditional.

Histone modification represents another layer of epigenetic control. DNA is wrapped around histone proteins, forming chromatin. When chromatin is tightly condensed, gene expression is suppressed. When it is relaxed, genes become accessible. Chemical modifications to histones, including acetylation and deacetylation, determine this accessibility. Longevity-promoting behaviors such as physical activity, fasting, and temperature variation increase histone acetylation in genes related to stress resistance, mitochondrial function, and repair. Conversely, chronic inflammation and insulin resistance promote histone configurations that favor aging pathways.

Non-coding RNAs add a further regulatory dimension. These molecules do not produce proteins but influence which genes are translated and how efficiently. MicroRNAs, for example, can suppress entire gene networks involved in metabolism, immune regulation, and cellular survival. Aging alters non-coding RNA profiles in ways that amplify inflammatory signaling and reduce regenerative capacity. Emerging research shows that exercise, sleep normalization, and dietary polyphenols can favorably shift non-coding RNA expression, reinforcing longevity pathways at the post-transcriptional level.

Epigenetic regulation is also deeply intertwined with circadian biology. Nearly every cell in the body operates on a molecular clock governed by gene expression cycles. Disruption of circadian rhythms—through irregular sleep, artificial light exposure, or erratic eating patterns—alters epigenetic marks on clock genes and downstream metabolic genes. This disruption accelerates insulin resistance, inflammation, and oxidative stress, all of which feed back into epigenetic aging. Conversely, restoring circadian alignment stabilizes gene expression and improves biological age markers.

Stress hormones exert a particularly strong epigenetic influence. Cortisol binds to receptors that directly interact with DNA, modifying gene expression related to immune function, inflammation, and energy allocation. Acute stress activates adaptive pathways, but chronic stress leads to persistent epigenetic reprogramming that favors short-term survival over long-term maintenance. Over time, this pattern suppresses repair mechanisms and accelerates tissue aging. Stress management is therefore not a psychological luxury but an epigenetic necessity.

Environmental exposures further shape epigenetic switches. Heavy metals, endocrine-disrupting chemicals, air pollution, and persistent organic pollutants alter DNA methylation and histone structure in ways that persist long after exposure ends. These epigenetic scars accumulate silently, often without immediate symptoms. Reducing toxic load and supporting detoxification pathways can gradually normalize epigenetic signaling, particularly when combined with nutritional support and metabolic health.

The critical insight of this chapter is that epigenetic switches operate cumulatively and contextually. No single factor determines aging trajectory. Instead, gene expression reflects the net balance of signals received over time. When pro-longevity signals dominate, repair and resilience remain active. When aging signals dominate, degeneration accelerates. This balance is adjustable.

Understanding these mechanisms transforms longevity from a vague aspiration into a structured practice. Each intervention discussed in this book—movement, fasting, sleep, nutrition, stress regulation—functions by flipping epigenetic switches toward maintenance rather than decay. The body does not resist aging out of stubbornness; it responds logically to the information it receives.

Epigenetic longevity is therefore not about overpowering biology, but about cooperating with it. When signals are consistent, rhythmic, and supportive, the genome shifts into a preservation mode. Aging slows not because time stops, but because damage accumulation slows and repair efficiency improves. This is the foundation upon which long-term healthspan is built.

Maria didn't chase perfection

Maria had spent most of her adult life doing everything right—or so she believed. She ate reasonably well, exercised when time allowed, and followed medical advice closely. Yet by her early fifties, her body seemed to be betraying her. Weight crept upward despite unchanged habits. Sleep fractured. Joint pain appeared without injury. Brain fog replaced mental sharpness. Doctors adjusted medications and reassured her that this was simply aging.

What Maria didn't realize was that her life had become epigenetically noisy. Long work hours disrupted her circadian rhythm. Chronic stress kept cortisol elevated. Meals were rushed and irregular. Exercise was sporadic. None of these factors were catastrophic alone, but together they sent conflicting signals to her genome day after day. Her genes adapted—not for longevity, but for survival under perceived threat.

Her turning point came not from a diagnosis, but from curiosity. Learning about epigenetic clocks unsettled her. The idea that her biological age could be moving faster than her calendar age reframed everything. Aging was no longer passive. It was responsive. That realization sparked something deeper than motivation—it sparked responsibility.

Maria didn't chase perfection. She focused on coherence. She anchored her sleep and meals to consistent times. She moved daily, not aggressively, but intentionally. She practiced stress reduction not to relax, but to recalibrate her biology. Slowly, imperceptibly at first, her body responded. Inflammation softened. Energy stabilized. Mental clarity returned.

Months later, blood markers improved. More importantly, Maria felt synchronized again. Her body stopped fighting itself. The aches didn't vanish overnight, but recovery improved. She no longer felt like she was aging faster than she could keep up with. She felt aligned.

What Maria learned was not how to stop aging, but how to stop accelerating it. By reducing epigenetic noise, she allowed her genome to reestablish order. The switches that once favored inflammation and fatigue began shifting toward repair and resilience.

Her story reflects a quiet truth about longevity. Most people do not age because they fail dramatically. They age because small misalignments accumulate unnoticed. When those misalignments are corrected, the body does not resist—it responds gratefully.

Maria's life did not become simpler. It became clearer. She understood that every day was a vote. Not for youth or age, but for signaling. Over time, those votes shaped her biology more reliably than genetics ever could.

Epigenetic switches are not flipped once. They are reinforced daily. And when reinforced consistently, they change the direction of aging itself.

Chapter 3

Rewriting Your Epigenetic Timeline

Aging is commonly perceived as a linear process, advancing steadily year after year. Epigenetic science reveals a different reality. Biological aging occurs in bursts, plateaus, and accelerations, shaped by periods of physiological stress, environmental exposure, and metabolic instability. These fluctuations form what can be described as an epigenetic timeline—a record of how your genome has responded to life's signals over time. Understanding this timeline is essential because it explains why aging can suddenly feel faster, and why strategic interventions can slow it dramatically.

Epigenetic clocks measure biological age by analyzing DNA methylation patterns across thousands of genomic sites. These clocks do not simply reflect time passed; they reflect how cells have been instructed to behave. Periods of chronic stress, illness, poor sleep, inflammation, or metabolic dysfunction leave identifiable epigenetic signatures. These signatures can persist long after the original trigger has resolved, effectively locking the body into an accelerated aging state unless corrective signals are introduced. One of the most critical insights from epigenetic research is that aging speed is adjustable. While chronological age advances uniformly, epigenetic age responds to signal quality and consistency. Individuals who experience prolonged psychological stress, circadian disruption, or inflammatory burden often show sudden jumps in biological age. Conversely, individuals who implement structured lifestyle changes can stabilize or even reduce biological age markers within months. Aging, therefore, is not continuous—it is negotiable.

The body interprets prolonged instability as a survival threat. In response, it reallocates energy away from long-term repair toward immediate function. This trade-off suppresses autophagy, DNA repair, mitochondrial renewal, and stem cell maintenance. Over time, these suppressed pathways create the visible and invisible markers of aging. Rewriting the epigenetic timeline requires reversing this energy prioritization, signaling safety rather than threat at the cellular level.

Consistency plays a more powerful role in epigenetic regulation than intensity. Extreme interventions applied sporadically often fail to produce lasting change because they do not alter baseline signaling expectations. In contrast, moderate but consistent behaviors—regular sleep timing, predictable feeding windows, daily movement, and controlled stress exposure—gradually reshape gene expression. The genome adapts to what it expects to encounter repeatedly. When signals stabilize, repair pathways regain priority. Midlife represents a critical inflection point in the epigenetic timeline. Around the fourth and fifth decades of life, accumulated signaling errors often converge. Hormonal shifts, metabolic changes, reduced physical activity, and increased psychological stress combine to accelerate epigenetic aging. This acceleration is not inevitable. It reflects a mismatch between biological design and environmental input. Correcting that mismatch during this period produces disproportionately large benefits, effectively bending the aging curve.

Rewriting the epigenetic timeline also involves removing persistent aging signals. Chronic inflammation, insulin resistance, sleep deprivation, and environmental toxicity continually reinforce aging pathways. Addressing these factors does not require perfection, but it does require intentional reduction. Each removed stressor reduces epigenetic noise, allowing clarity to return to gene regulation.

Importantly, epigenetic recovery is not instantaneous. The genome responds gradually as new patterns establish dominance. Early changes may not be immediately perceptible, which leads many individuals to abandon interventions prematurely. Understanding that epigenetic change operates on weeks-to-months timelines helps align expectations with biology. Longevity is a process of accumulation, not a breakthrough event. This chapter reinforces a foundational truth: your biological age reflects your signaling history, not your destiny. By changing what your cells experience consistently, you change how they behave. Over time, this rewrites the internal narrative of aging itself.

He standardized his routines

When Thomas turned fifty-two, the shift was sudden. Not dramatic, but unmistakable. His endurance dropped. Recovery lagged. Sleep lost depth. He felt as though his body had crossed an invisible threshold overnight. Friends reassured him that this was normal. Doctors echoed the sentiment. But Thomas sensed something more specific had changed—something directional.

Looking back, the signs had been accumulating for years. Long hours at work. Erratic meals. Chronic low-level stress. Late nights followed by early mornings. None of it felt extreme, but together they formed a pattern. His body had been adapting quietly, prioritizing survival over longevity. The sudden decline wasn't sudden at all—it was delayed feedback.

Thomas encountered the concept of biological age while listening to a lecture on epigenetics. What struck him wasn't fear, but clarity. Aging wasn't happening evenly. It was accelerating. And acceleration, he learned, could be slowed. He didn't attempt to reverse time. He focused on stabilizing it. He standardized his routines. Sleep became non-negotiable. Meals became rhythmic. Movement became daily. Stress became managed rather than ignored. He wasn't chasing youth—he was restoring predictability.

Months passed. Changes emerged gradually. Energy returned first. Then clarity. Then resilience. Injuries healed faster. Fatigue no longer lingered. Friends noticed the difference before he did. Thomas had not become younger. He had become biologically calmer.

What Thomas rewrote was not his calendar age, but his epigenetic trajectory. His body no longer perceived constant threat. Repair systems reactivated. Inflammation quieted. His genome adapted to the new expectations he set.

Years later, Thomas described aging differently. Not as decline, but as slope. When unmanaged, the slope steepened. When guided, it flattened. His body responded not to force, but to consistency.

His story illustrates the essence of epigenetic longevity. Aging speed is not fixed. It reflects accumulated instruction. When instruction changes, outcomes follow.

The timeline can be rewritten—not by erasing the past, but by redefining the present.

Chapter 4

Eliminating Epigenetic Noise

Aging does not accelerate solely because damaging signals increase. It accelerates because meaningful signals become harder for the genome to interpret. This confusion, known as epigenetic noise, disrupts the precision with which genes regulate repair, metabolism, and resilience. When epigenetic noise rises, cells lose clarity. Gene expression becomes erratic. Repair pathways misfire. Inflammation persists even in the absence of injury. The body begins aging not from damage alone, but from disorganization.

Epigenetic noise refers to the accumulation of conflicting, excessive, or poorly timed biological signals that interfere with coherent gene expression. These signals originate from many sources: irregular sleep, inconsistent eating patterns, chronic stress, blood sugar instability, environmental toxins, sedentary behavior, excessive stimulation, and persistent inflammation. Individually, these factors may appear benign. Collectively, they create a signaling environment so chaotic that the genome cannot maintain stable regulatory patterns.

In youth, epigenetic signaling is highly ordered. DNA methylation patterns are precise, histone modifications are well coordinated, and circadian gene expression follows predictable rhythms. With age and environmental disruption, this order deteriorates. Methylation becomes less targeted. Histone regulation loosens. Gene expression becomes increasingly stochastic. This loss of precision is not merely a marker of aging—it is a driver of it.

One of the most damaging forms of epigenetic noise arises from circadian disruption. When sleep timing, light exposure, and feeding windows vary unpredictably, clock genes lose synchronization. Because circadian genes regulate thousands of downstream pathways, their disruption cascades across metabolism, immune function, hormone release, and cellular repair. Over time, this desynchronization amplifies inflammatory signaling and suppresses regeneration, accelerating biological aging independent of calorie intake or exercise volume.

Metabolic instability is another powerful source of epigenetic noise. Frequent blood sugar spikes, prolonged insulin elevation, and erratic energy availability alter methylation patterns associated with inflammation and aging. Cells adapt by prioritizing short-term glucose management over long-term repair. This shift suppresses autophagy, mitochondrial renewal, and DNA repair mechanisms. The result is accelerated epigenetic aging even in individuals who appear outwardly healthy.

Chronic psychological stress contributes uniquely to epigenetic noise because it operates continuously and invisibly. Elevated cortisol alters gene expression directly by binding to DNA-associated receptors. Over time, stress-induced epigenetic marks persist, maintaining inflammatory and immune-suppressive states long after the original stressor has passed. This creates a feedback loop in which the body remains biologically vigilant, sacrificing longevity for perceived survival.

Environmental toxins further distort epigenetic signaling. Endocrine disruptors, heavy metals, air pollutants, and synthetic chemicals alter DNA methylation and histone structure in ways that persist across cell divisions. These changes do not always produce immediate symptoms, making them particularly insidious. Over years, they increase epigenetic entropy—the gradual loss of regulatory fidelity that defines biological aging.

Eliminating epigenetic noise does not require eliminating all stress or exposure. It requires reducing unnecessary variability and restoring rhythmic signaling. Predictability is biologically stabilizing. When sleep occurs at consistent times, meals follow reliable windows, movement is regular, and recovery is

longer did. She recovered faster. Her body felt organized again. What changed wasn't her ambition, but her signal quality. By removing noise, she allowed her genome to reestablish rhythm. Repair systems resumed. Inflammation softened. Her biology stopped bracing for impact.

Eleanor realized that aging hadn't been accelerating because she was doing too little—but because she was doing too much without rhythm. When she restored structure, her body responded with gratitude.

Her story illustrates a quiet truth of longevity. The body does not need constant improvement. It needs coherence. When signals align, aging slows naturally.

Epigenetic clarity is not loud. It is calm. And in that calm, longevity takes root.

Chapter 5

Nutritional Signals That Reprogram Longevity Genes

Food does far more than provide energy or structural material for the body. Every nutrient consumed carries biochemical information that interacts directly with gene regulation systems. Macronutrients influence hormonal signaling, micronutrients serve as cofactors for enzymatic processes, and bioactive compounds alter transcription pathways. Through these mechanisms, nutrition functions as one of the most powerful epigenetic modulators available in daily life. The genome continuously interprets dietary inputs to determine whether resources are abundant or scarce, stable or erratic, protective or inflammatory. These interpretations shape metabolic efficiency, repair capacity, immune balance, and ultimately the rate of biological aging.

One of the most direct nutritional influences on gene expression occurs through methylation pathways. DNA methylation requires methyl donors such as folate, choline, betaine, and methionine, along with cofactors including vitamins B2, B6, and B12. When these nutrients are sufficient and metabolically balanced, methylation patterns remain stable and targeted. When they are deficient or imbalanced, methylation becomes dysregulated, altering gene expression in ways associated with inflammation, metabolic dysfunction, and accelerated aging. Nutritional adequacy in these pathways therefore supports genomic stability across the lifespan.

Caloric intake itself carries epigenetic meaning independent of nutrient composition. Chronic caloric excess signals abundance, activating pathways related to growth and storage while suppressing maintenance processes such as autophagy and cellular repair. In contrast, moderate caloric restriction or intermittent fasting signals scarcity, triggering adaptive gene programs associated with stress resistance, mitochondrial renewal, and metabolic flexibility. These responses evolved to preserve function during periods of limited food availability. Modern longevity nutrition leverages these ancient pathways deliberately, using controlled nutritional stress to activate protective gene expression without inducing malnutrition.

Macronutrient balance further shapes epigenetic signaling through hormonal modulation. High glycemic carbohydrate intake elevates insulin repeatedly, promoting gene expression patterns linked to fat storage and inflammation. Excessive saturated fat combined with caloric surplus may amplify inflammatory gene pathways, particularly when metabolic health is compromised. Conversely, balanced protein intake supports muscle maintenance and metabolic regulation, while unsaturated fats and fiber-rich carbohydrates modulate gene expression toward anti-inflammatory and cardiometabolic resilience. The epigenetic impact of macronutrients therefore depends not only on quantity but on metabolic context.

Bioactive compounds in plant foods exert particularly strong epigenetic effects. Polyphenols such as resveratrol, quercetin, catechins, and curcumin influence histone acetylation and DNA methylation patterns associated with antioxidant defense, inflammation control, and mitochondrial function. These compounds do not act as nutrients in the traditional sense; rather, they function as signaling molecules that inform the genome about environmental conditions. Diets rich in diverse plant compounds provide a wide array of epigenetic cues that collectively reinforce longevity pathways.

The timing of nutrient intake also influences gene expression. Circadian biology coordinates metabolic gene activity with daily light-dark cycles. When feeding occurs during the biological day, metabolic pathways align with optimal hormone sensitivity and enzymatic activity. Late-night eating disrupts this alignment, altering epigenetic marks on circadian genes and impairing glucose regulation. Over time, irregular feeding schedules increase epigenetic noise and accelerate metabolic aging independent of total calorie intake. Consistent meal timing therefore supports genomic rhythm and stability.

Protein quantity and distribution provide another important epigenetic signal. Adequate protein maintains muscle mass and metabolic rate, supporting longevity. However, chronic excess protein, particularly from certain amino acid profiles, may activate growth pathways that suppress autophagy and repair. Longevity-focused nutrition often incorporates periods of moderate protein intake balanced with intermittent lower-protein intervals, mimicking natural cycles of abundance and scarcity. This pattern supports both tissue maintenance and regenerative signaling.

Inflammatory dietary components disrupt epigenetic regulation by activating immune-related gene networks persistently. Highly processed foods, refined sugars, and oxidized fats promote methylation changes associated with inflammatory gene expression. Over time, these patterns become entrenched, sustaining low-grade inflammation even when diet improves. Reducing inflammatory dietary inputs gradually normalizes these epigenetic marks, demonstrating the reversible nature of nutrition-driven gene expression.

Micronutrient density remains essential for enzymatic processes underlying epigenetic maintenance. Minerals such as zinc, magnesium, selenium, and iron support DNA repair enzymes, antioxidant systems, and methylation reactions. Deficiencies impair these processes, increasing oxidative damage and epigenetic drift. Nutritional adequacy therefore functions as a stabilizing force against aging-related genomic disorganization.

Importantly, nutritional epigenetics emphasizes pattern rather than perfection. Occasional dietary variability does not disrupt gene regulation. Chronic inconsistency does. Stable nutrient intake, balanced macronutrients, plant compound diversity, and rhythmic meal timing collectively provide coherent signals that reinforce longevity gene expression. Over time, these signals accumulate into durable biological advantages.

This chapter reframes nutrition as communication rather than consumption. Each meal conveys information about environmental conditions and resource availability. The genome interprets this information to allocate energy between growth, storage, and repair. When dietary signals align with metabolic stability and periodic challenge, longevity pathways dominate. Aging slows not through restriction alone, but through clarity of nutritional messaging.

The Language of Food

Samuel never thought of food as information. It was comfort, routine, culture, reward. Meals punctuated his day, often eaten quickly between obligations. He wasn't careless—just unintentional. By his late forties, fatigue lingered. Weight crept upward. Blood markers edged toward concern. Nothing dramatic, yet nothing optimal. Doctors recommended general improvements. Samuel complied intermittently, then returned to habit.

What shifted his perspective was a simple realization: his body wasn't reacting to single meals—it was reacting to patterns. The idea that nutrition spoke to his genes reframed everything. Food wasn't just fuel. It was instruction.

Samuel began observing rather than restricting. He noticed how irregular eating left him drained. How late meals disrupted sleep. How processed foods dulled energy. Instead of chasing diets, he stabilized rhythm. Meals occurred at consistent times. Whole foods replaced packaged ones gradually. Plants diversified his plate. Portions adjusted naturally as appetite recalibrated.

Weeks passed before changes felt tangible. Then energy steadied. Cravings softened. Sleep improved. He felt nourished rather than full. His body responded not to deprivation, but to clarity.

Months later, medical markers improved. More importantly, Samuel trusted his body again. Hunger felt honest. Satiety felt reliable. He no longer battled food. He collaborated with it. What Samuel learned was subtle but profound. Nutrition wasn't about discipline—it was about dialogue. His genome responded to repeated messages of stability and adequacy. Inflammation quieted. Metabolism recalibrated. Aging signals softened.

He didn't become perfect. He became consistent. Meals varied, but patterns held. Food regained meaning beyond taste or calories. It became communication.

Years later, Samuel described aging differently. Not as accumulation of years, but as accumulation of signals. When signals aligned, his body thrived. When they conflicted, decline accelerated. Understanding this gave him agency without obsession.

His story reflects the core principle of epigenetic nutrition. The body does not age from occasional indulgence. It ages from persistent miscommunication. When dietary signals become coherent, gene expression stabilizes, and longevity pathways reemerge naturally.

Food had always been speaking. Samuel finally learned how to listen—and how to answer.

Chapter 6

Movement Signals That Restore Youthful Gene Expression

Human physiology evolved under conditions of frequent physical activity interspersed with rest and recovery. Movement was not optional; it was the primary environmental signal informing the genome about survival demands, resource availability, and functional necessity. In modern environments, this foundational signal has diminished dramatically. Sedentary behavior has become normalized, and with it, the epigenetic programs that support metabolic efficiency, mitochondrial renewal, and tissue maintenance have weakened. Movement, therefore, is not merely exercise—it is a core regulatory input that preserves youthful gene expression patterns across the lifespan.

At the molecular level, physical activity alters DNA methylation, histone configuration, and gene transcription in tissues throughout the body. Skeletal muscle, adipose tissue, vascular endothelium, and even the brain exhibit epigenetic changes in response to movement. Genes associated with mitochondrial biogenesis, antioxidant defense, glucose transport, and lipid metabolism become upregulated with regular activity. Conversely, sedentary conditions promote methylation patterns linked to insulin resistance, inflammation, and metabolic dysfunction. These changes accumulate gradually, explaining why inactivity accelerates biological aging independent of body weight or diet quality.

Muscle tissue plays a central role in epigenetic longevity because it functions as both a metabolic regulator and a signaling organ. Contracting muscle releases myokines—signaling molecules that influence gene expression in distant tissues, including liver, fat, and brain. These signals enhance insulin sensitivity, reduce inflammation, and promote cellular repair pathways. Loss of muscle mass with age therefore represents more than structural decline; it reflects diminished systemic signaling for resilience and maintenance. Preserving muscle through regular movement maintains this endocrine-like communication network.

Mitochondrial health is particularly sensitive to movement signals. Physical activity stimulates mitochondrial biogenesis through activation of transcription factors such as PGC-1α, which regulates genes involved in energy production and oxidative stress defense. Increased mitochondrial density improves metabolic flexibility and reduces reactive oxygen species accumulation, both of which slow epigenetic aging. In contrast, sedentary behavior suppresses mitochondrial renewal, leading to energy inefficiency and increased oxidative damage. Movement thus sustains cellular energy systems essential for longevity.

The epigenetic effects of movement depend more on frequency and consistency than on intensity alone. Moderate, repeated activity produces stable gene expression changes that accumulate over time. Occasional high-intensity exercise without consistent baseline movement yields less durable epigenetic adaptation. The genome responds to repeated mechanical and metabolic signals by establishing new regulatory baselines. Daily movement therefore reinforces longevity pathways more effectively than sporadic exertion.

Movement also influences inflammatory gene regulation. Regular physical activity reduces expression of pro-inflammatory cytokine genes while increasing anti-inflammatory signaling. This shift occurs partly through changes in immune cell epigenetic profiles. Active individuals exhibit methylation patterns associated with reduced chronic inflammation and improved immune surveillance. Because inflammation is a primary driver of epigenetic aging, movement functions as a powerful anti-aging regulator at the genomic level.

Neurological gene expression responds similarly. Exercise increases neurotrophic factors such as BDNF, supporting synaptic plasticity and neuronal survival. These effects involve histone acetylation and transcriptional activation within brain regions responsible for memory and mood regulation. Sedentary behavior, in contrast, suppresses these pathways, contributing to cognitive decline and neurodegenerative risk. Movement therefore preserves brain longevity through epigenetic mechanisms parallel to those affecting metabolic health.

Mechanical loading of bone and connective tissue provides additional epigenetic signals. Weight-bearing activity stimulates gene expression involved in bone remodeling and collagen synthesis, maintaining structural integrity with age. Absence of loading leads to downregulation of these pathways, accelerating musculoskeletal aging. This illustrates a broader principle: tissues maintain function only when their mechanical demands are met consistently.

Circadian biology intersects with movement signaling as well. Physical activity performed during daylight hours reinforces circadian gene expression and metabolic timing. Sedentary lifestyles combined with irregular activity disrupt this alignment, increasing epigenetic noise. Regular movement at predictable times stabilizes rhythmic gene regulation across systems, amplifying longevity effects beyond the activity itself. Importantly, movement need not be extreme to exert epigenetic influence. Walking, resistance training, mobility work, and moderate aerobic activity all produce beneficial gene expression changes when performed consistently. The genome responds to mechanical engagement and energy demand, not to athletic achievement. Longevity movement emphasizes continuity rather than performance.

This chapter reframes movement as biological instruction rather than caloric expenditure. Muscles, mitochondria, immune cells, and neurons interpret mechanical and metabolic signals as indicators of environmental necessity. When movement signals persist, the genome maintains repair and resilience pathways. When they vanish, those pathways fade. Aging accelerates not because movement stops abruptly, but because signaling gradually diminishes.

Restoring movement restores information. And when information aligns with biological design, youthful gene expression patterns persist longer than chronological age alone would predict.

The Body Remembers Motion

Richard had always considered himself active. In youth, he played sports, lifted weights, moved constantly without thinking. But adulthood brought stillness disguised as responsibility. Work required sitting. Commutes lengthened.

Exercise became occasional rather than inherent. By his late fifties, Richard didn't feel inactive—just slower. Stiffer. Less capable than memory suggested.

The change had been gradual enough to seem inevitable. He told himself this was aging. Everyone slowed eventually. Yet the gap between how he remembered his body and how it felt widened each year.

A minor injury changed everything. A strained back from lifting something routine took weeks to resolve. The recovery felt disproportionate. His body seemed fragile in ways it never had before. Frustration replaced resignation. He began questioning whether decline was truly unavoidable.

Learning that movement influenced gene expression unsettled him. The idea that inactivity had been silently instructing his body to age reframed decades of habit. His muscles hadn't simply weakened; their signals had faded. His body hadn't forgotten motion—it had adapted to its absence.

Richard didn't begin intensely. He walked daily, at first reluctantly. Then consistently. Mobility work followed. Light resistance training returned slowly. Nothing dramatic changed in weeks. Yet months later, subtle differences emerged. Stiffness eased. Balance improved. Recovery shortened. His body responded as though recognizing something familiar.

What surprised him most was not strength returning, but confidence. He moved differently—not cautiously, but intentionally. His body felt engaged again, as though dormant systems had awakened.

Years later, Richard described aging differently. Not as loss, but as neglect reversed. Motion had always been his body's language. When he stopped speaking it, decline accelerated. When he resumed, resilience returned.

His story illustrates a quiet principle of epigenetic longevity. The body remembers motion even after years of absence. Genes once activated by movement do not vanish; they wait. When signals return consistently, expression resumes.

Richard never regained youth. He regained function. And with function came vitality he thought lost to time.

Movement did not make him younger. It reminded his genome how not to age faster than necessary.

The body remembers motion. And when motion returns, longevity follows.

Chapter 7

Stress Signals and the Genetics of Resilience

Stress is not inherently damaging. It is a biological signal that has shaped survival across evolutionary time. Acute stress activates adaptive pathways that enhance alertness, energy mobilization, immune readiness, and repair capacity. Problems arise not from stress itself but from chronic, unresolved activation of stress-response systems. When stress becomes persistent rather than episodic, the epigenetic landscape shifts. Genes that should activate briefly remain engaged continuously, altering inflammation, metabolism, and cellular maintenance in ways that accelerate aging.

The central mediator of the stress response is cortisol, released through activation of the hypothalamic-pituitary-adrenal axis. Cortisol binds to receptors that directly interact with DNA, influencing gene transcription in immune cells, metabolic tissues, and the brain. In acute situations, this signaling is protective, mobilizing energy and suppressing unnecessary processes temporarily. With chronic exposure, however, sustained cortisol alters DNA methylation and histone structure in genes associated with inflammation, insulin resistance, and immune dysregulation. These epigenetic changes can persist long after the original stressor resolves.

Chronic psychological stress is particularly impactful because it lacks physical resolution. Unlike exercise or environmental stressors that end predictably, psychological stress often remains cognitively active even in the absence of immediate threat. The body interprets persistent rumination or emotional strain as ongoing danger, maintaining stress gene expression patterns. Over time, this suppresses repair pathways such as autophagy and DNA maintenance, redirecting cellular resources toward vigilance rather than regeneration.

Stress also disrupts circadian gene expression. Elevated evening cortisol delays melatonin release, impairing sleep onset and quality. Poor sleep further increases cortisol, creating a reinforcing loop of circadian and epigenetic disruption. Because many repair processes occur during deep sleep, chronic sleep disturbance compounds the epigenetic effects of stress, accelerating biological aging through both hormonal and transcriptional pathways.

Inflammation represents another interface between stress and epigenetic aging. Persistent stress signaling activates immune pathways even in the absence of infection or injury. Pro-inflammatory cytokine genes become upregulated through methylation changes and transcription factor activation. This low-grade inflammation contributes to metabolic dysfunction, vascular aging, and tissue degradation. Stress-induced inflammation therefore acts as both a consequence and driver of epigenetic aging.

Recovery interrupts this cascade by restoring physiological safety signals. Parasympathetic nervous system activation counterbalances stress hormones, reducing cortisol and inflammatory gene expression. Practices that promote parasympathetic dominance—deep breathing, relaxation, social connection, nature exposure, and restorative sleep—produce measurable epigenetic effects. These include reduced methylation of stress-related genes and increased expression of repair-associated pathways.

Resilience emerges when stress exposure and recovery remain balanced. Moderate, temporary stress followed by adequate recovery strengthens adaptive gene expression, improving metabolic flexibility and immune competence. This pattern mirrors hormesis, in which controlled challenge enhances biological robustness. Without recovery, however, hormetic stress becomes chronic strain, accelerating aging instead of slowing it.

Epigenetic resilience therefore depends not on eliminating stress but on ensuring resolution. The genome expects fluctuations between activation and restoration. When these cycles occur reliably, stress-response genes reset appropriately. When activation persists without resolution, gene expression drifts toward chronic defense states incompatible with longevity.

Importantly, perception influences epigenetic stress responses. The brain interprets context before initiating hormonal signaling. Situations perceived as controllable or meaningful produce different gene expression patterns than those perceived as helpless or threatening. Psychological framing therefore alters biological impact. This explains why identical stressors affect individuals differently at the molecular level.

This chapter reframes stress management as genetic regulation rather than emotional comfort. Recovery signals inform the genome that threat has passed, allowing repair systems to resume priority. Over time, consistent recovery stabilizes gene expression, reduces inflammation, and slows biological aging.

Longevity does not require a stress-free life. It requires a life in which stress resolves.

The Return to Calm

Elena had always been capable. She handled responsibility efficiently, moved quickly, and adapted under pressure. Stress felt normal, even productive. Deadlines sharpened focus. Demands fueled momentum. For years, she thrived in acceleration.

Then something changed. Sleep fragmented. Fatigue lingered. Irritability surfaced unexpectedly. Her body felt tense even at rest. She attributed it to workload and pushed harder. The harder she pushed, the worse she felt. Doctors found nothing abnormal. "You're just stressed," they said. Elena didn't know how to be anything else.

The concept that stress altered gene expression unsettled her deeply. She had always viewed stress as mental, temporary, manageable. Learning that persistent stress could reprogram her biology reframed everything. Her body wasn't merely tired—it was adapting to continuous threat.

Elena began noticing how rarely she felt safe. Even in quiet moments, her mind rehearsed tasks and worries. Her body remained braced. She realized she had lost the ability to disengage. Stress had become baseline rather than response.

She started small. Breathing slowly before sleep. Walking without devices. Pausing between tasks. None felt dramatic. Yet they introduced something unfamiliar: stillness. Her body resisted at first, as if calm were unnatural. Gradually, it softened. Weeks passed. Sleep deepened slightly. Tension eased intermittently. She noticed brief moments of relaxation she hadn't felt in years. Encouraged, she protected recovery intentionally. Evenings slowed. Work boundaries formed. Rest became deliberate rather than accidental.

Months later, Elena felt different. Not less capable, but more stable. Stress still occurred, but it no longer lingered indefinitely. Her body recovered instead of remaining activated. Energy returned—not as urgency, but as steadiness.

What changed was not her environment but her signaling. By restoring calm, she told her genome that threat had ended. Repair resumed. Inflammation quieted. Her biology shifted from defense to maintenance.

Elena realized she had not been aging faster because life was hard. She had been aging faster because stress never resolved. When recovery returned, her body responded with relief.

Years later, Elena described resilience differently. Not as endurance, but as recovery capacity. Strength wasn't how much stress she could bear—it was how fully she could return to calm.

Her story reflects a central truth of epigenetic longevity. The body tolerates stress when it ends. It deteriorates when it persists. Calm is not absence of challenge—it is completion of it.

Elena did not remove stress from her life. She restored its rhythm. And in that rhythm, aging slowed.

Chapter 8

Environmental Signals That Shape Your Genetic Future

Human genes evolved within specific environmental conditions characterized by natural light cycles, clean air, mineral-rich water, diverse microbiological exposure, and relatively low chemical burden. These environmental inputs shaped gene regulation systems over thousands of generations. Modern environments differ dramatically from these ancestral conditions, introducing artificial light, air pollution, synthetic chemicals, processed materials, and indoor-dominated lifestyles. These changes alter epigenetic signaling in ways that influence inflammation, metabolism, detoxification, and cellular repair. Longevity therefore depends not only on internal habits but on the external conditions in which gene expression occurs.

Environmental signals influence epigenetics through direct and indirect mechanisms. Direct mechanisms include chemical interactions with DNA methylation pathways and histone modification enzymes. Indirect mechanisms operate through oxidative stress, endocrine disruption, immune activation, and circadian disturbance. Both pathways converge on gene regulation, shaping biological aging independently of genetics or lifestyle behaviors alone. This explains why individuals with similar habits may age differently when environmental exposures diverge.

Air quality exerts one of the most pervasive epigenetic effects. Particulate matter, combustion byproducts, and industrial pollutants generate oxidative stress that alters methylation patterns in genes associated with inflammation and cardiovascular function. Chronic exposure promotes persistent immune activation and endothelial dysfunction, accelerating vascular aging. These epigenetic changes can occur even at exposure levels considered moderate, accumulating gradually over years without obvious symptoms. Improving air quality exposure, through ventilation, filtration, and outdoor environmental selection, reduces these signals and supports genomic stability.

Chemical exposures further shape gene expression through endocrine disruption. Synthetic compounds found in plastics, pesticides, personal care products, and industrial materials can mimic or block hormonal signaling pathways. Because hormones regulate gene transcription extensively, interference alters epigenetic regulation in metabolic, reproductive, and neurological systems. Some of these changes persist across cell generations, illustrating how environmental exposures can leave durable epigenetic imprints. Reducing unnecessary chemical contact lowers this interference and restores hormonal signaling clarity.

Light exposure represents another powerful environmental regulator of gene expression. Circadian genes synchronize cellular repair, hormone release, metabolism, and immune function with the day-night cycle. Natural daylight reinforces this rhythm, while artificial light at night disrupts it. Exposure to blue-spectrum light in evening hours suppresses melatonin and alters methylation patterns in clock genes, impairing sleep and recovery processes. Over time, circadian disruption increases inflammation and metabolic dysfunction through epigenetic pathways. Aligning light exposure with natural cycles therefore stabilizes gene regulation across systems.

Water quality contributes to epigenetic health through mineral content and contaminant load. Trace minerals such as magnesium, zinc, and selenium support enzymatic processes involved in DNA repair and methylation. Conversely, contaminants including heavy metals and industrial residues disrupt these pathways and increase oxidative stress. Chronic exposure to such contaminants alters gene expression patterns linked to inflammation and neurological aging. Ensuring clean, mineral-balanced hydration supports genomic maintenance indirectly but meaningfully. The built environment also influences epigenetic signaling through physical and sensory conditions. Indoor spaces often limit microbial diversity, natural light, and environmental variability. Lack of microbial exposure may impair immune gene regulation, increasing inflammatory sensitivity. Insufficient daylight weakens circadian alignment. Limited temperature variation reduces adaptive stress signaling. These factors collectively create a biologically monotonous environment that diminishes resilience pathways. Introducing natural elements, airflow, light variability, and environmental diversity restores ancestral signaling patterns.

Noise exposure contributes less obviously but meaningfully to epigenetic aging. Persistent background noise elevates stress hormones and sympathetic activation, altering gene expression associated with inflammation and cardiovascular strain. Even when consciously ignored, noise maintains physiological vigilance. Quiet environments support parasympathetic activity and recovery signaling, reinforcing genomic stability. Environmental calm therefore functions as a biological regulator rather than merely a psychological preference.

Temperature exposure similarly affects gene expression. Mild cold and heat exposures activate adaptive pathways related to mitochondrial function, vascular health, and stress resistance. Chronic thermoneutral environments, in contrast, reduce these signals. While extreme exposure is unnecessary, periodic environmental variation provides hormetic input that reinforces resilience gene networks. This illustrates how environmental complexity supports longevity at the epigenetic level.

Importantly, environmental epigenetics emphasizes cumulative exposure rather than single events. The genome responds to repeated conditions over time. Small improvements in air, light, chemical load, and environmental rhythm accumulate into meaningful longevity advantages. Conversely, chronic low-level exposures accumulate into epigenetic burden. Environmental quality therefore functions as a background signal shaping aging continuously.

This chapter reframes environment as biological instruction rather than passive context. Genes interpret surroundings to determine threat, safety, abundance, and rhythm. When environmental signals align with evolutionary expectations, repair and resilience pathways dominate. When signals conflict with biological design, aging accelerates silently. Longevity requires not only internal discipline but external alignment.

Your environment is speaking to your genes constantly. And over years, it helps determine how quickly or slowly you age.

The Place That Changed Her Biology

Nora never questioned her environment. She lived in a modern apartment, worked indoors, and spent most days under artificial light. Air felt normal. Water tasted fine. Noise blended into background familiarity. She focused on diet and exercise, believing health came solely from personal behavior. Yet fatigue lingered, sleep felt shallow, and recovery lagged despite her efforts.

Her turning point came unexpectedly during a brief retreat in a rural setting. Mornings began with sunlight. Air felt cleaner, quieter. Evenings darkened naturally. Meals remained similar to home, activity levels unchanged. Yet within days, Nora felt different. Sleep deepened. Mood lifted. Energy stabilized. The shift felt disproportionate to the minor change in location.

Returning home, the contrast became obvious. Indoor air felt heavy. Nights glowed with artificial light. Noise hummed constantly. She realized her environment had been shaping her biology silently. Her body had adapted to conditions she never noticed until they changed.

Nora began adjusting her surroundings intentionally. Morning sunlight exposure became routine. Windows opened for airflow. Evening light softened. Noise reduced where possible. Water filtration improved. Indoor plants and natural materials increased environmental texture. None of these changes were dramatic, yet together they altered how her space felt—and how her body responded.

Weeks later, sleep improved noticeably. Fatigue decreased. Calm returned more easily. She hadn't changed discipline or diet significantly; she had changed context. Her genes responded not only to her actions but to her surroundings.

Nora realized health was never solely internal. The body listens to environment continuously, interpreting air, light, sound, and chemical exposure as biological instruction. When these signals aligned with natural patterns, her body relaxed into repair rather than vigilance.

Years later, Nora described that retreat not as escape, but as revelation. Her biology had not improved because life slowed. It improved because environmental signals clarified. She carried that clarity home, reshaping her daily surroundings to match what her body recognized as safe and restorative.

Her story reflects a quiet principle of epigenetic longevity. The environment does not merely host life—it directs it. When surroundings support rhythm and safety, gene expression stabilizes. Aging slows without force. Nora did not change who she was. She changed where she lived biologically. And in that shift, her body found resilience it had been missing.

Your environment is never neutral. It is always shaping your genetic future. Nora simply learned how to shape it back.

Chapter 9

Protecting the Epigenetic Youth of the Brain

The brain ages differently from other organs, yet it remains deeply influenced by epigenetic regulation. Neurons are largely non-dividing cells, meaning they must maintain function across decades without replacement. Their longevity depends on precise gene expression controlling synaptic plasticity, energy metabolism, inflammation regulation, and repair. When epigenetic stability persists, cognitive clarity, memory, and emotional balance remain resilient. When epigenetic drift accumulates, neuronal signaling deteriorates, accelerating cognitive aging and neurodegenerative risk.

One of the primary drivers of brain epigenetic aging is neuroinflammation. Microglial cells, the immune regulators of the brain, become increasingly reactive with age and environmental stress. Chronic activation alters DNA methylation patterns in genes associated with inflammatory signaling and synaptic maintenance. Over time, this shifts the neural environment toward sustained low-grade inflammation, impairing communication between neurons and reducing plasticity. Factors such as poor sleep, metabolic dysfunction, psychological stress, and environmental toxins amplify this process through epigenetic pathways.

Mitochondrial function plays a critical role in neuronal longevity. The brain consumes a disproportionate share of the body's energy, relying heavily on efficient mitochondrial activity. Epigenetic regulation controls genes involved in mitochondrial biogenesis and antioxidant defense within neurons. Physical inactivity, chronic stress, and metabolic disease suppress these pathways, reducing energy availability and increasing oxidative damage. Conversely, movement, metabolic flexibility, and intermittent energetic challenge enhance mitochondrial gene expression, preserving neuronal vitality.

Synaptic plasticity—the brain's ability to form and strengthen connections—depends on epigenetic modulation of transcription factors and neurotrophic proteins. Brain-derived neurotrophic factor supports neuronal growth, repair, and learning capacity. Epigenetic modifications such as histone acetylation increase BDNF expression in response to movement, cognitive engagement, and enriched environments. Reduced stimulation or chronic stress decreases these modifications, lowering neurotrophic support and accelerating cognitive decline. The brain therefore requires ongoing environmental and behavioral signals to maintain adaptive gene expression. Sleep exerts profound epigenetic influence on brain aging. During deep sleep, neuronal repair pathways activate, metabolic waste clears, and synaptic homeostasis resets. Sleep deprivation alters methylation patterns in genes governing circadian rhythm, inflammation, and synaptic function. Chronic disruption impairs memory consolidation and increases neurodegenerative risk. Restorative sleep restores epigenetic regulation across neural networks, emphasizing its central role in cognitive longevity.

Metabolic health intersects with brain epigenetics through glucose regulation and insulin signaling. Neurons rely on stable energy supply, and insulin resistance impairs this availability. Elevated glucose and insulin alter methylation patterns in neuronal and vascular genes, increasing oxidative stress and inflammation. Over time, these changes contribute to cognitive decline. Nutritional stability and metabolic flexibility therefore protect brain gene expression by ensuring consistent energy signaling.

Psychological stress influences neuronal epigenetics through cortisol-mediated transcription changes. Prolonged stress reduces expression of genes supporting synaptic plasticity while increasing inflammatory gene activity. These effects can persist, affecting mood and cognition long after stressors resolve. Recovery and stress resolution restore balanced gene expression, highlighting the brain's sensitivity to emotional environment at the molecular level.

Environmental enrichment supports brain epigenetic resilience. Novelty, learning, social interaction, and sensory stimulation activate gene networks associated with plasticity and repair. These experiences increase histone acetylation and neurotrophic signaling, strengthening neural circuits. Deprivation of stimulation produces the opposite effect, reducing gene expression linked to cognitive maintenance. The brain requires engagement to sustain youthful epigenetic patterns.

Importantly, neuronal epigenetic aging is modifiable throughout life. Even in later decades, targeted interventions—movement, cognitive challenge, sleep restoration, stress regulation, and metabolic balance—can alter gene expression favorably. Neurons retain plasticity longer than once believed, provided signaling conditions support adaptation. Cognitive aging therefore reflects not only chronological time but cumulative epigenetic environment.

This chapter reframes brain aging as a process of regulatory drift rather than inevitable decline. Neuronal genes respond continuously to metabolic, behavioral, and environmental inputs. When signals align with stability and engagement, plasticity persists. When signals degrade, neuronal resilience fades. Protecting the epigenetic youth of the brain requires sustaining the inputs that neurons evolved to expect: movement, learning, rhythm, and recovery.

The brain does not age solely because years pass. It ages according to the signals it receives repeatedly. And those signals remain adjustable across the lifespan.

The Mind That Stayed Bright

Arthur feared losing his mind more than his body. In his sixties, he began noticing small lapses—names slipping, words pausing, thoughts trailing. They were subtle, common, easily dismissed. Yet they unsettled him deeply. His father had declined cognitively in later years, and Arthur worried he was following the same path.

He maintained physical health diligently, but cognitive vitality felt less controllable. Memory seemed fragile, dependent on forces beyond effort. He assumed decline was inevitable, something to endure rather than influence.

Learning that brain aging reflected gene expression changed everything. The idea that neuronal resilience depended on ongoing signals reframed his fear. His mind wasn't fading randomly—it was responding to environment and habit. That realization restored agency. Arthur began introducing deliberate cognitive engagement. He learned new skills slowly. Read complex material. Practiced recall intentionally. Movement increased, sleep improved, stress reduced. None of these changes were extreme. Yet together they created stimulation his brain had lacked.

Months passed. Lapses decreased. Focus strengthened. Conversations flowed more easily. He felt mentally sharper, not younger but clearer. His confidence returned alongside function.

Arthur realized he had been protecting his brain too cautiously, avoiding challenge out of fear. In doing so, he had reduced stimulation his neurons required. When he reintroduced engagement, plasticity returned.

Years later, Arthur described aging differently. Not as fading light, but as adjustable brightness. His mind reflected how actively he used it and how well he supported its biology. Decline was not predetermined—it was directional.

His story reflects the central truth of brain epigenetics. Neurons remain responsive when stimulated consistently. Gene expression supporting plasticity persists when the brain is engaged, rested, and metabolically stable. When stimulation fades, so does resilience. Arthur did not prevent aging. He influenced its course. His mind stayed bright not by chance, but by signal.

The brain listens continuously to how it is used. Arthur simply changed what it heard.

Chapter 10

Integrating the Signals That Slow Biological Aging

Epigenetic longevity does not emerge from isolated interventions. It arises from the integration of signals that collectively inform the genome about environmental stability, resource availability, and survival demands. Throughout this book, individual influences—nutrition, movement, stress, recovery, environment, cognitive engagement—have been examined separately to clarify their mechanisms. In biological reality, however, these signals operate simultaneously, interacting continuously within the epigenetic regulatory network. Aging trajectory therefore reflects not any single behavior, but the coherence of all inputs combined.

The genome interprets cumulative signaling patterns to determine energy allocation between maintenance and immediate function. When signals suggest safety, rhythm, and metabolic stability, repair pathways dominate. Autophagy remains active, DNA repair persists, mitochondrial turnover continues, and inflammatory tone remains low. When signals suggest instability, unpredictability, or chronic threat, resources shift toward short-term survival, suppressing maintenance functions. Over time, this shift manifests as accelerated biological aging. Integration of longevity signals ensures that the genome receives a consistent message favoring preservation rather than defense.

Signal coherence is more influential than signal intensity. Extreme interventions applied inconsistently create biological contradiction. For example, vigorous exercise paired with chronic sleep deprivation and metabolic instability produces mixed epigenetic instructions. Similarly, optimal nutrition combined with persistent psychological stress fails to sustain repair dominance. The genome responds to overall context rather than isolated excellence. Longevity therefore depends on alignment across domains rather than perfection within any one domain.

Rhythm provides the framework through which integration occurs. Circadian biology synchronizes gene expression across systems, coordinating metabolism, immune activity, hormone release, and cellular repair with environmental cycles. When sleep timing, feeding windows, activity patterns, and light exposure align consistently, gene regulation stabilizes across tissues. This stability reduces epigenetic noise and preserves regulatory precision. Disruption of rhythm fragments signaling coherence even when individual behaviors remain beneficial. Integration therefore begins with temporal consistency. Energy signaling also requires integration.

Nutritional intake, metabolic health, and movement collectively determine cellular perception of energy availability. Excessive intake without expenditure signals storage, suppressing repair. Expenditure without adequate nutrition signals scarcity, activating stress pathways. Balanced cycles of intake and expenditure signal adaptability, promoting mitochondrial renewal and metabolic flexibility. Longevity emerges when energy signals alternate predictably between challenge and recovery rather than remaining chronically elevated or depleted.

Stress and recovery integration further determines epigenetic trajectory. Acute stress followed by resolution strengthens adaptive gene expression. Chronic stress without recovery maintains inflammatory signaling and suppresses repair. Recovery practices—sleep, parasympathetic activation, psychological disengagement—complete the stress cycle, restoring genomic equilibrium. Without recovery, even beneficial challenges become aging accelerants. Integration therefore requires ensuring that activation phases conclude with restoration.

Environmental alignment integrates external signals with internal regulation. Light exposure, air quality, chemical load, and sensory conditions continuously influence circadian and inflammatory gene expression. When environmental inputs support rhythm and safety, internal interventions amplify their effects. When environment contradicts internal behavior—such as healthy diet within polluted or disrupted surroundings—epigenetic benefits diminish. Longevity therefore depends on congruence between behavior and environment.

Cognitive engagement integrates neurological signals with systemic physiology. Learning, novelty, and mental challenge activate neurotrophic gene expression that interacts with metabolic and inflammatory pathways. The brain's epigenetic state influences hormonal regulation, stress response, and behavioral adherence. Sustained cognitive stimulation therefore reinforces whole-body longevity signaling. Integration recognizes the brain not as isolated but as regulatory hub.

Importantly, integration simplifies rather than complicates longevity practice. When signals align naturally, fewer interventions are required. Regular sleep supports metabolic health. Consistent movement enhances stress resilience. Stable nutrition reinforces circadian rhythm. Each domain amplifies others, creating positive epigenetic feedback. Fragmented interventions require greater effort because signals conflict. Integrated living reduces biological friction.

Epigenetic aging reflects the net clarity of biological instruction over time. Disjointed signals produce regulatory drift. Coherent signals maintain precision. Integration ensures that all major domains—energy, rhythm, stress, environment, cognition—convey the same message: conditions are stable, predictable, and safe for long-term maintenance. Under such conditions, the genome favors preservation.

This final chapter reframes longevity as orchestration rather than optimization. The body does not require extreme biohacking to age slowly. It requires consistent, aligned signaling across fundamental domains of life. When alignment persists, biological age progression decelerates naturally.

Longevity is not achieved through any single action repeated intensely. It is achieved through many signals repeated coherently. And coherence is the true language of youthful gene expression.

The Pattern That Changed His Life

Victor spent years searching for the perfect longevity strategy. He tried diets, supplements, training programs, sleep trackers, cognitive apps—each pursued intensely for a time, then abandoned when results plateaued. Progress always felt temporary. He improved in one area while another declined. Health felt like a moving target rather than a stable state. He believed longevity required optimization. More data, more precision, more intervention. Yet the more he pursued perfection, the more fragmented his life became. Sleep shifted around workouts. Diet changed with trends. Stress remained constant. Effort increased without coherence.

The realization that aging reflected integrated signaling unsettled him. He had been improving variables individually while ignoring their interaction. His body wasn't confused by lack of effort—it was confused by inconsistency.

Victor stopped optimizing and began aligning. Sleep became fixed. Meals stabilized. Movement became daily rather than episodic. Stress recovery was protected. Environment simplified. Nothing extreme changed, yet everything coordinated. For the first time, his routines supported each other rather than competed.

Months passed. Changes felt subtle but durable. Energy stabilized. Mood steadied. Recovery improved. Health stopped fluctuating dramatically. He no longer chased results; they accumulated naturally.

Victor recognized that his body had not needed perfection—it needed clarity. When signals aligned across domains, his genome shifted toward maintenance without force. Aging did not reverse, but it slowed into something manageable. Years later, Victor described longevity differently. Not as a collection of habits, but as a pattern. When the pattern held, resilience persisted. When it fractured, decline accelerated. His focus moved from maximizing individual behaviors to sustaining overall coherence.

His story reflects the core principle of epigenetic integration. The body does not respond to isolated excellence. It responds to consistent alignment. When nutrition, movement, sleep, stress, and environment convey the same message repeatedly, gene expression stabilizes in a longevity-favoring state.

Victor never found a perfect protocol. He found a coherent life. And in that coherence, his biology aged more slowly than it otherwise would have.

Longevity did not come from doing more. It came from signals working together.

The pattern changed. And so did his future.

Chapter 11

Designing Your Epigenetic Future

Epigenetic longevity ultimately extends beyond slowing aging in the present; it involves shaping how aging unfolds across the decades ahead. Gene expression does not merely reflect current conditions—it also establishes regulatory momentum. Patterns reinforced today influence how cells respond tomorrow. This creates an epigenetic trajectory, a directional aging pathway formed through accumulated signaling over time. Designing an epigenetic future therefore requires sustaining conditions that reinforce stability, adaptability, and repair long enough for them to become the biological default.

The genome continuously predicts future demands based on past patterns. When environments remain consistent, cells optimize for efficiency. When conditions fluctuate unpredictably, cells prioritize vigilance and survival. Long-term longevity depends on convincing the genome that stability will persist. This allows maintenance systems to remain active rather than intermittently suppressed. Epigenetic future design therefore centers on establishing enduring patterns rather than temporary interventions. One of the most influential determinants of future epigenetic aging is cumulative inflammatory burden. Even when inflammation appears low at any single moment, repeated activation across years produces durable methylation changes in immune and metabolic genes. These marks bias tissues toward inflammatory responses later in life, increasing disease susceptibility. Sustained reduction of inflammatory triggers—metabolic instability, stress, environmental toxins, and circadian disruption—gradually reshapes immune gene expression toward tolerance rather than reactivity. Over years, this alters the baseline inflammatory set point.

Metabolic flexibility also determines long-term epigenetic trajectory. Cells capable of transitioning efficiently between fuel sources maintain mitochondrial health and oxidative balance. Chronic metabolic rigidity, characterized by persistent glucose dependence and insulin elevation, alters gene expression toward storage and inflammation. Sustained patterns of nutritional rhythm, periodic energy challenge, and movement preserve metabolic adaptability, ensuring that energy regulation genes remain responsive rather than fixed in aging-accelerating states.

Hormonal signaling stability contributes similarly. Hormones regulate transcription across nearly every tissue. Chronic disruption—whether through stress, circadian misalignment, or endocrine interference—creates persistent epigenetic alterations in metabolic, reproductive, and neurological genes.

Restoring rhythmic hormonal signaling through sleep consistency, light alignment, and stress resolution gradually recalibrates these pathways. Over time, endocrine gene expression returns to patterns associated with youthful regulation.

Neuronal epigenetic health influences long-term trajectory through behavior reinforcement. Cognitive clarity supports adherence to longevity-promoting patterns. Cognitive decline disrupts routine stability, indirectly accelerating aging. Sustained mental engagement, learning, and emotional regulation preserve neuronal plasticity, which in turn supports behavioral continuity. This feedback loop illustrates how brain epigenetics governs whole-body aging trajectory indirectly.

Environmental exposure patterns also accumulate into long-term genomic imprinting. Chronic exposure to pollutants, chemical disruptors, or circadian disruption produces methylation marks that persist across cell divisions. Reducing these exposures consistently over years allows gradual epigenetic normalization. While some marks persist, overall regulatory noise declines, improving genomic precision in aging tissues. Environmental consistency therefore shapes future biological aging as much as present function. Recovery capacity represents another long-range determinant. The ability to resolve stress efficiently preserves repair pathways across decades. When recovery remains intact, stress-induced epigenetic marks reset rather than accumulate. When recovery weakens, stress signatures compound. Practices that maintain parasympathetic balance, sleep quality, and psychological regulation therefore protect the genome from long-term stress imprinting. Longevity depends less on avoiding stress than on ensuring its resolution repeatedly across life.

Importantly, epigenetic trajectory changes gradually but persistently. Improvements sustained for months begin to alter baseline gene expression. Improvements sustained for years redefine regulatory norms. The genome adapts not to aspiration but to repetition. Designing an epigenetic future therefore requires maintaining coherent signaling long enough for stability to become biological expectation rather than temporary condition.

This perspective reframes aging from passive inevitability to directional process. While chronological time advances unavoidably, the biological rate and quality of aging remain influenced by cumulative instruction. Future cellular behavior reflects present signaling continuity. By sustaining aligned patterns, individuals influence how tissues function decades ahead. Epigenetic longevity ultimately depends on persistence. Short interventions produce transient benefits. Long-term coherence reshapes trajectory. When metabolic, hormonal, neurological, and environmental signals remain stable across years, gene expression drifts toward preservation rather than decline. Aging does not stop, but it unfolds more slowly and with greater functional integrity.

Designing your epigenetic future therefore means committing not to perfection, but to continuity. The genome responds to what is repeated. And what is repeated long enough becomes destiny at the cellular level.

The Years He Gave Back to Himself

When Leonard turned sixty, he began thinking about time differently. Not the time already lived, but the time remaining. His health was decent, his habits reasonable, his life stable. Yet he sensed a quiet acceleration. Recovery took longer. Energy fluctuated more. Aging felt less like gradual passage and more like downward slope.

He wondered how much of that slope was fixed. His parents had aged quickly in later years. He assumed he would follow. Genetics felt like gravity—inescapable.

The idea that aging trajectory could be shaped unsettled him. He had always thought of health in present terms: how he felt today, how he functioned this year. Considering decades ahead required different thinking. It meant acknowledging that patterns now would define outcomes later.

Leonard didn't pursue dramatic change. He chose continuity. Sleep became consistent across seasons. Movement became daily regardless of mood. Nutrition stabilized. Stress resolution became intentional. Environment simplified. He removed variability rather than adding complexity.

Months passed with modest change. Then years passed. Subtle differences accumulated. His peers aged visibly—energy fading, mobility declining, illness increasing. Leonard noticed he was diverging. Not younger, but steadier. Aging had not stopped, yet its pace differed.

At seventy, the divergence became unmistakable. He recovered faster than friends. Moved more easily. Thought clearly. His trajectory had bent. Not dramatically, but persistently. He had not defied aging; he had reshaped its slope.

Leonard realized he had not changed genetics—he had changed expectation. His genome adapted to decades of stable signals. Repair remained active longer. Inflammation stayed lower. Resilience persisted.

He often reflected on the quiet power of repetition. None of his habits were extraordinary. Their persistence was. While others fluctuated between effort and neglect, his pattern held. Over years, that pattern became biology.

Leonard did not regain youth. He extended function. The years he feared losing to decline remained usable, active, present. He had given them back to himself through continuity.

His story illustrates the essence of epigenetic future design. Aging trajectory responds to sustained signaling. What repeats becomes regulatory baseline. When stability persists long enough, cells expect preservation rather than deterioration.

Leonard never stopped aging. He simply aged more slowly than he would have otherwise. And that difference was measured not in days, but in years of life lived well.

The future of the body is written gradually. Leonard learned to write it deliberately.

Epilogue

The Biology of Continuity

Aging is often perceived as the gradual loss of vitality, clarity, and resilience over time. Yet throughout this book, a different perspective has emerged. Aging is not solely the passage of years; it is the cumulative effect of signals interpreted by the genome. Each day, cells read environmental, behavioral, metabolic, and psychological information to determine whether conditions favor repair or survival. This interpretation shapes biological aging continuously, moment by moment, across decades.

Epigenetic longevity reveals that aging trajectory is not fixed. Gene expression patterns remain responsive throughout life, influenced by nutrition, movement, stress resolution, environment, sleep, and cognitive engagement. These inputs do not operate in isolation. They converge into integrated biological instruction. When signals align toward stability and safety, maintenance pathways persist. When signals fragment or contradict, repair yields to deterioration. Longevity therefore reflects coherence more than effort.

The most profound implication of epigenetic science is that biological age is adjustable. Not reversible in the sense of returning to youth, but modifiable in rate and quality.

Individuals who sustain aligned patterns of living often experience slower functional decline than expected for their chronological age. This is not genetic luck alone; it is the result of repeated signaling that preserves genomic regulation. Over time, stability becomes the body's expectation.

Importantly, epigenetic longevity does not demand perfection. Cells respond to dominant patterns rather than isolated deviations. Occasional disruption does not determine trajectory. Chronic incoherence does. Consistency in sleep, nutrition, movement, and recovery gradually stabilizes gene expression even when life remains complex. Longevity emerges from continuity, not control.

The body's capacity for adaptation persists longer than once believed. Even in later decades, improved signaling alters inflammatory tone, metabolic flexibility, mitochondrial function, and neuronal plasticity. These changes occur gradually but meaningfully. Aging does not cease, but its acceleration slows. Functional years expand. Healthspan extends into lifespan.

Throughout these chapters, a recurring principle has appeared: the genome listens continuously. It interprets light and darkness, activity and rest, abundance and scarcity, safety and threat. These interpretations become biology.

Understanding this transforms longevity from abstract aspiration into daily practice. Every behavior communicates. Every environment instructs. Every rhythm reinforces expectation.

Epigenetic longevity ultimately reframes responsibility. Health is neither fully predetermined nor entirely controllable. It is shaped through interaction between genetic potential and lived experience. While genes establish capacity, environment and behavior guide expression. Individuals cannot eliminate aging, but they can influence how it unfolds.

The significance of this perspective extends beyond personal health. It changes how aging itself is understood. Decline becomes less inevitable and more conditional. Prevention shifts from treatment of disease to preservation of regulation. Longevity becomes not merely survival, but sustained function and engagement.

This book has explored mechanisms and applications, but its central message remains simple. The body ages according to the signals it receives repeatedly. When those signals align with biological design, resilience persists. When they conflict, decline accelerates. Longevity is therefore less about doing more and more about doing consistently what supports coherence. The future of aging science will continue to expand understanding of gene regulation, cellular repair, and biological time. Yet the foundational truth revealed here is unlikely to change: stability supports preservation. Rhythm sustains repair. Integration maintains vitality. These principles transcend specific interventions.

As you close this book, remember that epigenetic longevity is not a program to complete but a pattern to sustain. Its power lies not in intensity but in repetition. The genome adapts gradually, reflecting what is practiced long enough. Over years, this adaptation becomes trajectory.

Aging remains inevitable. Accelerated aging is not. Through sustained alignment of behavior, environment, and recovery, biological decline slows into something more gradual, more functional, and more livable. The years gained are not added from outside; they are preserved from within.

Longevity, ultimately, is continuity expressed through biology.

The Life That Continued

When Margaret turned seventy-five, she did not feel young. She felt present. Her days moved with rhythm—waking with light, moving without urgency, eating simply, resting fully. Nothing about her routine seemed extraordinary. Yet those who knew her noticed something unusual: she remained engaged in life with clarity and steadiness uncommon among her peers. Margaret had not always lived this way. In earlier years, life had been hurried and irregular. Sleep varied, meals rushed, stress constant. Aging felt abrupt then—fatigue rising, resilience fading. She assumed this was inevitable. Everyone around her aged similarly.

Her change began quietly in her fifties. Not through crisis, but realization. She saw that her body responded predictably to patterns. When sleep stabilized, energy returned. When meals regularized, metabolism steadied. When stress resolved, calm lingered. These connections seemed obvious in hindsight, yet transformative in practice.

Margaret simplified gradually. Rhythm replaced variability. Movement became daily. Evenings slowed. Environment softened. She did not seek youth. She sought stability. Over years, that stability accumulated into biology.

At seventy-five, Margaret noticed she differed from friends. Not in age, but in slope. Their decline felt steeper. Hers felt gradual. She moved carefully but confidently. Thought clearly. Recovered reliably. Aging had not stopped; it had slowed.

She often reflected on the quiet power of repetition. None of her habits were dramatic. Their persistence was. While others alternated between effort and neglect, her pattern held. Over decades, that pattern shaped her trajectory.

Margaret did not believe she had extended life. She believed she had preserved function. The years she lived remained usable—filled with movement, thought, connection. Longevity, she realized, was not measured by years added but by years retained.

She aged as all do. But she aged on a gentler curve. And that difference changed everything.

www.ingramcontent.com/pod-product-compliance
Ingram Content Group UK Ltd.
Pitfield, Milton Keynes, MK11 3LW, UK
UKHW020157290926
15471UKWH00005B/926

9 798295 670992